BRITISH GEOLOGIC

# Geology of the Winchester district

a brief explanation of the geological map
Sheet 299 Winchester

K A Booth

© NERC copyright 2002

*Bibliographic reference*

BOOTH, K A. 2002. Geology of the Winchester district —
a brief explanation of the geological map.
*Sheet Explanation of the British Geological Survey.*
1:50 000 Sheet 299 Winchester (England and Wales).

Keyworth, Nottingham: British Geological Survey

# CONTENTS

1 **Introduction** 1
2 **Geological description** 4
   Devonian to Triassic 4
   Jurassic 4
   Cretaceous 7
   Palaeogene 16
   Quaternary 20
   Structure 23
3 **Applied geology** 26
   Hydrogeology 26
   Bulk minerals 27
   Hydrocarbons 27
   Geotechnical considerations 28
**Information sources** 30
**Selected bibliography** 32

## Figures

*Inside front cover*   Geological succession of the district
1 Structure of the Wessex–Weald Basin 1
2 Deep boreholes in the district 5
3 Concealed Jurassic strata 6
4 Correlation of boreholes 8–9
5 Concealed Cretaceous strata 11
6 Correlation of the Chalk 12
7 Structure of the district 24
8 Potential ground constraints 29
9 Aeromagnetic anomaly map 30
10 Bouguer gravity anomaly map 31

## Plates

1 Twyford M4 cutting 2
2 Newhaven Chalk Formation 3
3 King's Somborne Lime Quarry 16
4 Reading Formation sands 17
5 Reading Formation, clay and pebbles 18
6 Floodplain deposits of the River Test 22
7 Solution pipe 28

## Notes

The word 'district' refers to the area of the 1:50 000 Series geological map Sheet 299 Winchester. National grid references are given in square brackets; unless otherwise stated all lie within the 100 km square SU. Borehole records referred to in the text are prefixed by the code of the National Grid 25 km$^2$ area upon which the site falls, for example SU 42 SW, followed by its registration number in the BGS collection. Lithostratigraphical symbols shown in brackets in the text, for example (NCk) refer to those shown on the published 1:50 000 map. Numbers at the end of photograph descriptions refer to the official collection of the BGS.

## Acknowledgements

This Sheet Explanation was compiled by K A Booth. D J Evans contributed to the description of the concealed strata and C P Royles provided information on geophysics. The manuscript was edited by P M Hopson and A A Jackson. The author's thanks are due to the many landowners, tenants and quarry companies for access to land.

The grid, where it is used on figures, is the National Grid taken from Ordnance Survey mapping.

© Crown copyright reserved Ordnance Survey licence number GD272191/2002.

*Front cover*
Anthony Gormley's statue in the flooded Crypt, Winchester Cathedral (*Photograph* by David Fox, reproduced with the kind permission of the Dean and Chapter of Winchester Cathedral).

Subsidence problems at the cathedral have affected, in particular, the south-east chapel, where the gravel base on which the cathedral is built slopes towards the river and the water table is close to the surface.

The medieval foundations were inadequate, and this end of the cathedral began to collapse at the turn of the 20th century; a major restoration project was undertaken, but even today the crypt floods in the winter months.

# 1 Introduction

This *Sheet Explanation* provides a summary of the geology for the part of north-east Hampshire around Winchester and Stockbridge (Sheet 299). The district covers an area from Winchester and the M3 motorway in the east to the Hampshire – Wiltshire border (10 km east of Salisbury) in the west. It stretches from near Andover (and the A303 trunk road) in the north, to near Romsey in the south. The district is founded for the greater part on the Chalk (Plates 1; 2) with Palaeogene strata preserved in the south. The broad gently dipping slopes and short, steep scarps characteristic of the strata of the area, are cross-cut by the north–south-orientated Itchen and Test Valleys.

Structurally, the district falls within the Wessex Basin (Figure 1), which comprises a system of post-Variscan extensional sedimentary basins and 'highs' that covered much of southern England, south of the London Platform and Mendip Hills, during Permian to Mesozoic times. At greater depths are Palaeozoic strata which were strongly deformed during the Variscan Orogeny, a period of tectonic upheaval and mountain building that culminated at the end of the Carboniferous. The rocks of the 'Variscan Basement' are weakly metamorphosed sandstones and limestones of Devonian and Carboniferous age. Several major southward-dipping thrust zones and north-west orientated wrench faults, thought to have originated during the Variscan Orogeny, have been tentatively identified in the basement. This deformation was followed by a long period of erosion and a major unconformity marks the base of the Permo-Triassic sequence.

In Permian times, subsidence associated with periods of tectonic extension began to affect southern England, initiating the development of a number of smaller fault-bounded basins within the Wessex Basin. Sedimentation in the expanding Wessex Basin began to the west of this district. Deposition gradually spread eastwards, so that the earliest Mesozoic rocks present at

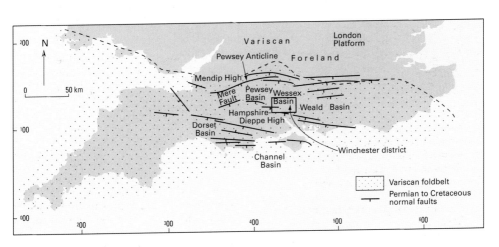

**Figure 1**   Structures of the Wessex–Weald Basin.

# 2 INTRODUCTION

**Plate 1** Twyford M4 cutting, Winchester [4895 2763]. New Pit Chalk and Lewes Nodular Chalk Formations, dipping 4° south are exposed in this northern section of the cutting (GS 1160).

depth within the district are a thin sequence of red beds thought to be of Triassic age. Crustal extension was accommodated by reactivation of existing faults in the Variscan basement, which show evidence of syndepositional downthrow to the south during Permian and Mesozoic times. The largest of these faults divides the region into a series of structural provinces (Chadwick, 1986) such as the Weald/Wessex and Channel Basins, separated by the Hampshire–Dieppe High (also known as the Cranborne–Fordingbridge High); the main Variscan front lies to the north of the district (Figure 9 ML; Busby and Smith, 2001).

This district straddles the northern margin of the Hampshire–Dieppe High and part of the Wessex Basin; the boundary between these two structural provinces lies along the Portsdown–Middleton Faults which underlie associated anticlines (Hopson, 2000a). The Hogs Back–Kingsclere–Pewsey (Figure 10 GL1) structure to the north of this district marks the boundary between the Weald Basin and the London Platform in the Andover–Basingstoke area.

Syndepositional movement on major faults resulted in thicker Jurassic sequences being laid down on the downthrown sides, the beds commonly thinning against the tilted fault blocks. Major periods of active extensional faulting occurred during the Jurassic and during deposition of the 'Wealden Group' of the Lower Cretaceous. During periods of tectonic quiescence, rates of sedimentation increased evenly towards the depocentre of the Weald Basin towards the east of this district.

The sea began to flood the Wessex Basin in Rhaetian (Late Triassic) times, depositing the

Penarth Group. The area of marine deposition increased gradually throughout the Jurassic, although minor periods of erosion occurred, mainly at the basin margins. By Upper Oxfordian to Kimmeridgian times, the London Platform was probably entirely submerged. Towards the end of Kimmeridgian times, the London Platform began to re-emerge, probably due to a combination of global sea-level fall and a reduction in the rate of tectonic subsidence. This resulted in erosion on the margins of the Wessex Basin and the beginning of the development of the Late Cimmerian unconformity. This marine regression continued into Cretaceous times, the environment of deposition changing from offshore marine (Kimmeridge Clay Formation) through shallow marine (Portland Group), to brackish water and evaporites (Purbeck Group), and fluviatile ('Wealden Group'). The final period of extensional fault movement, marked by normal faulting, resulted in the accumulation of thick sequences of 'Wealden Group' sediments in the main fault-bounded troughs in the Wessex Basin, whereas the intervening exposed highs suffered severe erosion.

A period of regional subsidence followed, and combined with eustatic rise in sea level, led to a renewed marine transgression of the Wessex Basin. The ensuing deposition of the Lower Greensand Group, Gault Formation, Upper Greensand Formation, and eventually the Chalk Group covered all the surrounding high areas, including the London Platform. A global fall in sea level at the end of the Cretaceous resulted in erosion of parts of the higher Chalk units and the development of a pre-Cainozoic unconformity. Later, deposition in Eocene to Oligocene times was followed by the onset of the compressive tectonic regime during mid-Neogene 'Alpine' earth movements. These movements effectively reversed the sense of movement on the major bounding faults of the Wessex and Channel basins causing inversion of the basins and highs. Uplift is estimated at about 1500 m (Simpson et al., 1989) for both the former Weald and Channel depocentres. Subsequently, erosion has unroofed these inverted basins giving rise to the present-day landscape.

Cross-sections showing the main structures are presented on Sheet 299 Winchester. The major extensional faults that control the asymmetric fold structures such as the Dean Hill Anticline can be seen beneath West Dean (Section 1, Sheet 299). The location of the Winchester–East Meon Pericline and associated fault can be noted (Section 2, Sheet 299) and the dramatic thinning of the Wealden and Lower Greensand Formations to the west (Section 1) is thought to be caused by progressive onlap of younger strata. The major folds are also revealed by the structure contours on the base of the Stockbridge Rock Member and Palaeogene deposits.

**Plate 2** Anastomosing marl seam within Newhaven Chalk Formation. The marl and the bed below are associated with sponges causing the yellow-brown colouration. King's Somborne Lime Quarry [SU 3380 2740] (GS 1086).

# 2 Geological description

The stratigraphy of the rocks buried beneath the district is known from boreholes sunk primarily for the hydrocarbon industry. Those at Farley (SU 22 NW/2) [2358 2852], Lockerley (SU 32 NW/15) [4306 1259], Goodworth (SU 34 SE/14) [3694 4195] and Furzedown (SU 32 NE/3) [4368 1284], together with a number of boreholes around Winchester (falling into sheets SU 42 NE, SU 53 SW and SU 52 NW) and Stockbridge (SU 43 NW, SU 43 NE, SU 43 SW, and SU 33 NE) form the basis of this account. The thicknesses of strata within these boreholes are summarised in Figure 2. At depth, a thin Permo-Triassic sequence of limestone, siltstone, sandstone and breccia overlies beds of siltstone and orthoquartzite tentatively assigned to the Devonian 'basement'. The structural contours and subcrops of the sub-Permian surface for the southern half of the United Kingdom (Smith, 1985) show a broad band of Devonian rocks stretching from south of the Mendips south-eastwards, across this district, and into part of the English Channel.

## DEVONIAN TO TRIASSIC

The Devonian has been penetrated in Goodworth 1 (SU34SE/14) [3694 4195] and Stockbridge 4 (SU33NE/3) [3964 3765] in the north of this district. Up to 40 m of 'Old Red Sandstone' was proved beneath a major unconformity that removed much of the Carboniferous strata beneath this district. These rocks are characterised predominantly by sandstone with subordinate siltstone and mudstone preserved in fining-upwards cycles of fluviatile origin.

These early **Devonian** continental sequences were deposited on the Brabant Massif to the north of the Cornwall Basin (Ziegler, 1982), a part of the Variscan Foredeep Basin; both basins derived much of their sediment from the continent of Laurasia to the north. **Carboniferous** strata above the post-Devonian unconformity have been encountered within boreholes to the south. These sequences are principally Carboniferous (Dinantian) limestone, and up to 280 m was preserved in Farley 1 in the far west of the district (SU22NW/2) [2358 2852].

A thick sequence of **Permo-Triassic** (P-T) strata is preserved above the unconformity throughout the district, namely Sherwood Sandstone Group, Mercia Mudstone Group and Penarth Group. These have a maximum proved thickness of 315 m in the Goodworth 1 Borehole (SU 34 SE/14) [3694 4195]. The Sherwood Sandstone Group is of variable thickness ranging from 10 m in the southwest to 57 m in the north. It consists of red, yellow and brown sandstone, pebbly in part with subordinate red mudstone and siltstone. The Mercia Mudstone Group consists of mottled reddish brown and greyish green calcareous siltstone and calcareous mudstone with some thin sandstone. The overlying Penarth Group comprises fissile, dark grey mudstone and white to pale brown limestone.

## JURASSIC

The whole of the Jurassic sequence is represented at depth below the district (Figure 3). They are mainly marine in origin and rest conformably on the Penarth Group having been deposited within the subsiding Wessex Basin. The relatively uniform, cyclical sequences of the Jurassic provide evidence, regionally, for an eastward shift of the area of maximum subsidence in the Wessex Basin when the faults bounding the Hampshire–Dieppe High became active. The Weald and Channel Basin depocentres developed at this time. In general, the beds thicken northwards into the Weald Basin against major faults.

**Figure 2** Summary of the principal deep boreholes in the Winchester district. Values are thicknesses of the deposit, given in metres.

| Borehole | Farley 1 | Goodworth 1 | Stockbridge 1 | Stockbridge 4 | Stockbridge 2 | Stockbridge 7 | Stockbridge 5 | Stockbridge 6 |
|---|---|---|---|---|---|---|---|---|
| Quarter Sheet | SU22 NW/2 | SU34SE/14 | SU43NE/5 | SU33NE/3 | SU43NE/8 | SU43SW/4 | SU43NW/7 | SU43NW/8 |
| Grid Reference | 4235 1285 | 4369 1419 | 4451 1355 | 4396 1376 | 4450 1355 | 4423 1338 | 4406 1357 | 4434 1364 |
| Upper Greensand F | 44.4 | 45.6 | 47.4 | 47 | 50 | 46 | 56 | 55 |
| Gault F | 51 | 58.9 | 64 | 59 | 68 | 62 | 54.4 | 61 |
| Lower Greensand F | 17.7 | 12.5 | 8.6 | 14.3 | 10 | 20 | 11.8 | 14 |
| Wealden 'Group' | absent | 183 | 237.4 | 219 | 252 | 224 | 235.3 | 268 |
| Purbeck Group | 28.3 | 62 | 51 | 96 | 75 | 88 | 106 | 123 |
| Portland Group | 50 | 55 | 61 | 86 | 73 | 83 | 90 | 102 |
| Kimmeridge Clay Formation | 184 | 218 | 234 | 384 | 438 | 348 | 434 | 472 |
| Corallian Group | 48 | 47.5 | 38 | 65 | 61 | 54 | 69 | 66 |
| Oxford Clay Formation | 151 | 141.5 | 135.5 | 201 | 221 | 168 | 215 | 211 |
| Kellaways Formation | 11 | 11 | 10 | 18 | 16 | 16 | 17 | 15 |
| Great Oolite Group | 118 | 122 | 118 | 157 | 352(?) | 151+ | 159+ | 177+ |
| Inferior Oolite Group | 41.5 | 85 | 77 | 105 | 73+ | | | |
| Lias Group | 256.5 | 487 | 433 | 528.5 | | | | |
| Permo-Triassic | 254 | 315 | 89 | 312 | | | | |
| Devonian-Carboniferous | 287+ | 28+ | 50+ | 40+ | | | | |

| Borehole | Furzedown | Winchester 1 | Winchester 2 | Winchester 3 | Winchester 4 | Winchester 5 | Lockerley |
|---|---|---|---|---|---|---|---|
| Quarter Sheet | SU32NE/3 | SU52NW/1 | SU52NW/2 | SU42NE/4 | SU53SW/1 | SU52NW/3 | SU32NW/15 |
| Grid Reference | 4368 1284 | 4503 1284 | 4544 1276 | 4470 1277 | 4510 1301 | 4502 1270 | 4306 1259 |
| Upper Greensand F | 47 | 40.5 | 39 | 43 | 33 | 41 | 41 |
| Gault F | 47 | 77 | 80 | 65 | 78 | 70 | 49 |
| Lower Greensand F | 8 | 25 | 26+ | 20 | 31 | 20 | 35 |
| Wealden 'Group' | 151 | 341 | | 290 | 254 | 290+ | not recorded |
| Purbeck Group | 60 | 77 | | 24+ | 18+ | | 39 |
| Portland Group | 54 | 80 | | | | | 55 |
| Kimmeridge Clay Formation | 222 | 275 | | | | | 186 |
| Corallian Group | 49 | 109 | | | | | 51 |
| Oxford Clay Formation | 147 | 146 | | | | | 148 |
| Kellaways Formation | 15 | 11 | | | | | 12 |
| Great Oolite Group | 118 | 131 | | | | | 122 |
| Inferior Oolite Group | 86 | 128 | | | | | 65 |
| Lias Group | 518 | 287+ | | | | | 421 |
| Permo-Triassic | 66+ | | | | | 412+ | |
| Devonian-Carboniferous | | | | | | | |

F  Formation

**Figure 3** The major subdivisions of the concealed Jurassic strata.

| Lithostratigraphical division | Thickness in metres | Map code | Divisions lithologies | Principal components | Subsidiary | Notes |
|---|---|---|---|---|---|---|
| Purbeck Group | 28-123 | Pb | Durlston and Lulworth formations | evaporites pass up to calcareous mudstone and shelly limestone | cherty, ooidal, fissile in parts | Erosional contact at base in places |
| Portland Group | 50-102 | Pl | Portland Stone and Portland Sand formations | sandstone and argillaceous sandstone pass up into shelly limestone | thin siltstones, mudstones, glauconitic | |
| Kimmeridge Clay Formation | 184-472 | KC | Upper, Middle and Lower | cycles of mudstone, 'oil-shale' and limestone | fissile and calcareous in part | |
| Corallian | 38-109 | Cr | Upper and Lower | limestone, sandstone | siltstone, mudstone | |
| Oxford Clay Formation | 135-221 | OxC | Weymouth, Stewartby and Peterborough | sandy, mudstones, locally calcareous | silty, calcareous, carbonaceous | |
| Kellaways Formation | 11-18 | Kys | Kellaways Sand and Clay members | mudstone passes up into fine-grained sandstone | micaceous, calcareous | Single, coarsening upwards sequence |
| Great Oolite Group | 118-177+ | GtO | Cornbrash, Forest Marble, Great Oolite and Fuller's Earth formations | pyritic siltstone and mudstone pass up into limestones | fissile, calcareous, passes up into ooidal, shelly packstone | Principal oil reservoir of Weald Basin |
| Inferior Oolite Group | 41-128 | InO | Upper, Middle and Lower | limestone and calcareous siltstone | sandy, ferruginous, becoming ooidal | |
| Lias Group | 256-572 | Li | Upper, Middle and Lower | interbedded mudstone and limestone | | Thickness varies greatly across district |

# CRETACEOUS

Lower Cretaceous strata in southern Britain comprise an important sequence that shows considerable vertical and lateral variation in both thickness and facies. It is generally fullest and thickest towards the basin centres (Whittaker et al., 1985). The Cretaceous period began with a short-lived marine transgression. Despite renewed subsidence at this time, clastic deposition in the Weald Basin was maintained in a nonmarine facies by an abundant sediment supply from the uprising London–Brabant Ridge to the north, Armonica to the south, and other land masses to the west and south-west. These lower Cretaceous sedimentary rocks are informally called the Wealden 'Group' and include the Hastings Group and the Weald Clay Formation.

## Wealden 'Group' (W)

The Wealden 'Group' is an informal term that includes the Weald Clay Formation and Hastings Group. Deposition was largely in a freshwater environment, in a large shallow lake or lagoon that occupied much of the present area of Hampshire and the Weald. Some of the major siltstone-sandstone bodies are thought to have formed by lateral accretion from migrating channels of braided streams, but the thickest sand units are attributed to accretion of sediment transported into the basin from a rejuvenated block-faulted source area. At this time the Channel and Weald basins are thought to have been separated by the 'Portsdown Swell' (the successor to the Hampshire–Dieppe High), and the Wealden 'Group' is known to thicken northwards away from this structure. Thicknesses vary greatly across the district and parts of the succession are absent in some boreholes.

In the Winchester district, the Lower Cretaceous strata are not exposed and relatively poorly understood with little published regarding their distribution and development. This is particularly so of the lowermost Wealden 'Group', which is generally concealed, beneath the Lower Greensand, Gault, Upper Greensand and Chalk of the district. However, a series of deep hydrocarbon exploration wells drilled across the Winchester and surrounding districts (Figure 4) provide valuable information on the subsurface nature and distribution of the concealed Wealden 'Group' and other Lower Cretaceous rocks. Geophysical logs were run through the Lower Cretaceous strata in a number of these boreholes and they show characteristic log signatures that can be widely correlated between boreholes and to lithostratigraphical units. The correlations of the Lower Cretaceous Wealden 'Group' presented here are based upon the published geophysical log interpretations of similar sequences encountered in the Collendean Farm Borehole in the Weald Basin (Whittaker et al., 1985).

Correlation diagrams of the Lower Cretaceous sequences (Figure 4) illustrate that considerable lateral variation exists in the individual units. This is seen not only in the thinning of units to the north and west, but also the gamma-ray and sonic log responses which reflect lithological variations. However, despite the considerable thinning of the Wealden 'Group', there appears to have been no northerly or westerly overstepping of the earlier Wealden 'Group' sequences by the succeeding sequences. This is illustrated by the fact that the main sandstones and mudstones, including the basal Fairlight Clay and the (thin) Lower Tunbridge Wells Sand, can be identified in boreholes beyond the district, for example Egbury [SU 4447 5236] to the north, and Farley South [SU 2358 2852] in the west, (Figure 4). The Welford Park Station Borehole [SU 4066 7364] proved Kimmeridge Clay overlain by thin Lower Greensand Group, illustrating that between it and the Egbury Borehole the entire Wealden, Purbeck, Portland and uppermost Kimmeridge Clay sequences are truncated beneath the Lower Greensand (Figure 4a).

The lateral change in log character is most clear in the mudstone of the Weald Clay

## 8 GEOLOGICAL DESCRIPTION

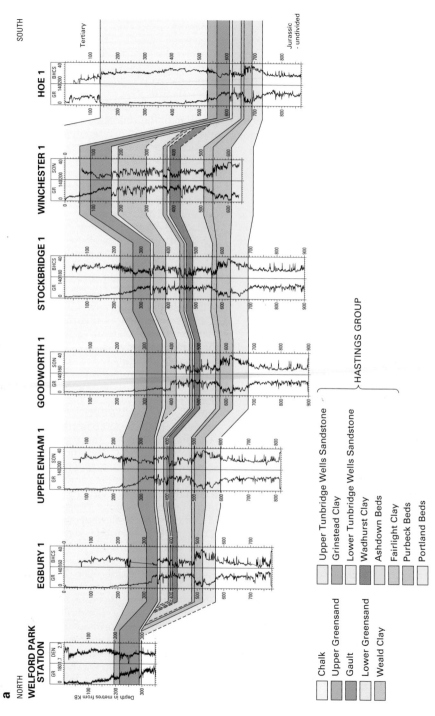

**Figure 4** Correlation of boreholes.

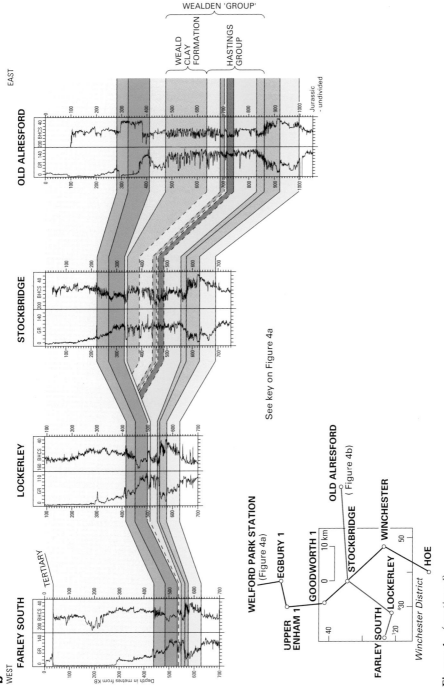

**Figure 4** (continued)

Formation and seems directly attributable to lateral facies/lithological variations. In thicker more basinward locations the gamma-ray response of the Weald Clay is high, the log signature being of a fine and highly serrated character. To the north and west, the signal becomes more deeply serrated ('ratty'), with increasingly thick units of lower gamma-ray values giving a blocky character. There are also indications of higher gamma-ray units decreasing gradually upwards in a cyclic nature. These responses are indicative of a passage to siltier and ultimately sandier facies, reflecting proximity to basin margins to the north and west of the district. It also serves to make the distinction between the Upper Tunbridge Wells sands and the sandier Weald Clay Formation increasingly difficult towards the basin margins.

The attenuation of the Wealden 'Group' towards the basin margins may occur partly through onlap. However, the dramatic thinning of the Weald Clay Formation is strongly suggestive of the removal of the younger Weald Clay succession beneath the unconformable Lower Greensand. Evidence of this is seen in borehole logs. Between Winchester, Stockbridge and Goodworth boreholes (Figure 4a), and more obviously the Old Alresford and Stockbridge boreholes (Figure 4b), the formation thins dramatically. Farther westwards, the Weald Clay Formation is absent in the Lockerley Borehole (Figure 4b).

## Lower Greensand Group (LGS)

Rising sea levels flooded the Wessex Basin leading to the re-establishment of marine conditions. The boundary between the non-marine sequence, the Wealden 'Group', and the overlying, marine Lower Greensand Group is marked by the Late Cimmerian unconformity. This represents a gap in the sequence that is greatest at the margins of the basin, where much of the lowest Cretaceous is missing. This gap diminishes towards the centre of the basin and the unconformity is identified only by a number of hiatuses or breaks in deposition. Tidally influenced shallow marine and shoreline sandstones and mudstones form the Lower Greensand succession. Thicker sequences were deposited in the Wessex Basin which subsided faster than the London–Brabant Ridge. Deepening of the basin continued into Albian times when the Gault Formation, a sequence of deeper marine mudstones, was deposited. The concealed Cretaceous strata are summarised in Figure 5.

Deep boreholes also reveal notable changes in the Lower Greensand Group to the north and west of the district. The dramatic westerly thinning of the group, seen on the Sheet 300 Alresford, where it is absent in the Strat B1 Borehole, continues across this district. Here, the Lower Greensand is either absent or much attenuated and below the resolution of the geophysical logs in the Farley 1 Borehole (SU22NW/2) [2358 2852]. However, there are departures from this overall pattern of thinning. For example, Welford Park Station Borehole [SU 4066 7364] to the north of the district proved 9.4 m of the Lower Greensand Group.

The nature of the thinning of the Lower Greensand Group is equivocal; it may thin by onlap against the basin margin, rather than by truncation beneath younger strata of Lower Cretaceous age.

## Gault Formation (G)

The Gault Formation consists mainly of pale to dark grey, fissured, soft mudstone with scattered phosphatic nodules up to 15 mm across. The thickness is relatively uniform across the district, thinning only slightly to the south-west. Its maximum thickness (up to 80 m) in this district was proved in Winchester 2 Borehole (SU 52 NW) [4544 1276].

## Upper Greensand Formation (UGS)

This formation consists of pale, yellow-brown, grey and greenish grey bioturbated siltstone and silty, very fine-grained sandstone, with variable amounts of mica and

**Figure 5** Concealed Cretaceous strata.

| Litho-stratigraphical division | Thickness in metres | Map code | Divisions lithologies | Principal lithologies | Subsidiary |
|---|---|---|---|---|---|
| West Melbury Marly Chalk Formation | 25–30 | WMCk | West Melbury Marly Chalk Fm with Glauconitic Marl Member | Highly glauconitic glauconitic sand overlain by grey marly chalk | Thin hard, grey limestones |
| Upper Greensand Formation | 33–56 | UGS | | Highly glauconitic calcareous silt and sandstone | Clay lenses and siltstones |
| Gault Formation | 47–80 | G | | Fissured silty clay | Scattered phosphatic nodules |
| Lower Greensand Group | 8–35 | LGS | Folkestone, Sandgate, Hythe and Atherfield Clay formations | Fossiliferous mudstone overlain by glauconitic medium-grained sandstone | Common siliceous sandstone and chert beds in Hythe Formation |
| Weald Group | 150–341 | W | Upper and Lower | Non-calcareous carbonaceous mudstone | Thin sandstones and sparse limestones |

glauconite. The thickness is relatively uniform, up to 56 m over much of the district but it thins rapidly south-eastwards.

## Chalk Group

Up to about 400 m of Upper Cretaceous chalk underlies the district, forming the extensive dip slopes that cross the area. The stratigraphical nomenclature used in this district is based on that of Mortimore (1986a) modified by Bristow et al. (1995, 1997) and as revised by the Geological Society Stratigraphic Commission (Rawson et al., 2001). The Chalk Group is divided into the Grey Chalk Subgroup and the White Chalk Subgroup, which are subdivided into ten formations that form the basis of lithostratigraphical mapping (Bristow et al., 1997) (Figure 6).

In Cenomanian times, landmasses were present in south-west England, Wales, Scotland and Northern Ireland, and farther afield in Brittany. Southern Britain lay approximately 10° of latitude farther south than its present position. Chalk accumulated on the outer shelf of an epicontinental subtropical sea of normal salinity and with little terrigenous input.

GREY CHALK SUBGROUP

This unit, between 85 and 90 m thick comprises two formations, the **West Melbury Marly Chalk Formation** overlain by the **Zig Zag Chalk Formation**. The lower formation is entirely concealed in this district but is known to be a repetitive sequence of hard limestone and softer marl (calcareous mudstone) couplets, each pair being between 0.5 and 2 m thick at most. The base of the formation is marked by the **Glauconitic Marl Member**, an arenaceous, glauconitic, marly sandstone which provides a distinctive positive gamma-ray peak in downhole geophysical logs throughout southern

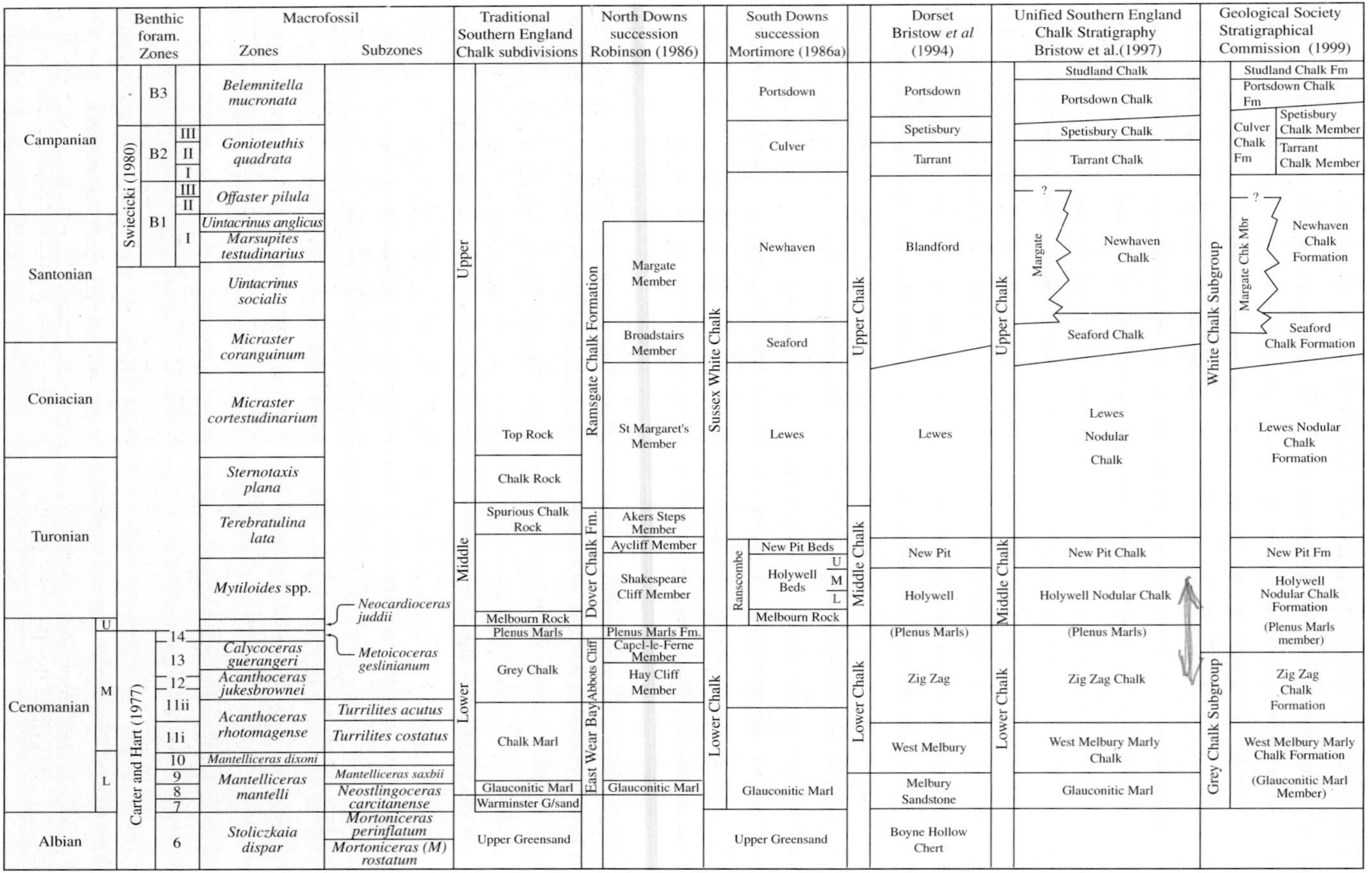

**Figure 6** Correlation of zonal schemes for the Chalk of southern England.

England. This formation is estimated to be 25 to 30 m thick at depth in this district.

### Zig Zag Chalk Formation (ZCk)

This formation is composed typically of medium hard, greyish white, blocky chalk but maintains in its lower part the limestone/marl couplet characteristics of the formation below. In this district the Zig Zag Chalk is between 55 and 60 m thick but only the uppermost 40 m is seen at outcrop. The base of the formation is the Cast Bed (Bristow et al., 1997), which, outside this district, is known to be a very fossiliferous, pale yellow-brown calcarenite. This marks a period of Mid-Cenomanian erosion which frequently removes the upper part of the West Melbury Marly Chalk Formation. However, in full sequences the Cast Bed rests on the Tenuis Limestone, the highest unit of the formation below.

Higher in the succession the formation becomes less marly and is pale cream or white in colour. This colour change is thought to occur at the level of 'Jukes-Browne Bed 7' (a calcarenite bed with phosphatic nodules). The top of the formation is taken as the base of the **Plenus Marls Member**.

The Zig Zag Chalk outcrops only in the east near Winchester. There are no extensive exposures in this area at the time of survey. Much of the outcrop is beneath pasture/recreation grounds and has been heavily landscaped during the construction of the M3 motorway (Hopson, 2001).

Jukes-Browne and Hill (1903) mentioned a pit 'recently opened at Bar End' which exposed 'firm, massive chalk, greyish when wet but drying white' which they attributed to the 'Holaster subglobosus' Zone on a fairly extensive fossil collection. During this survey a small degraded pit, partly infilled and heavily overgrown, was noted at [4889 2817] exposing Zig Zag Chalk.

### WHITE CHALK SUBGROUP

This unit comprises eight formations with only the highest part of the Portsdown and Studland Chalk Formations absent due to pre-Palaeogene erosion. It is characterised by white chalk with numerous flint seams. The combined thickness is up to 350 m.

### Holywell Nodular Chalk Formation (HCk)

The Holywell Nodular Chalk Formation includes the Plenus Marls Member, with a series of closely spaced, brightly coloured marl beds at its base. These are usually 1 to 2 m thick and are rarely exposed, although they can be identified in field brash at a number of localities. The marls are overlain by the hard nodular chalks of the Melbourn Rock Member (about 5 m thick), which generally lacks significant shell debris. The overlying chalk comprises medium-hard to very hard, nodular chalk with flaser marls throughout. The chalk is commonly shelly and has a gritty texture. The Holywell Nodular Chalk Formation is between 25 and 30 m thick, and is seen at outcrop around Winchester itself.

### New Pit Chalk Formation (NPCk)

The New Pit Chalk comprises medium-hard, massive-bedded, pure white chalk with regularly spaced pairs or groups of marls, each up to 15 cm thick. It is sparsely fossiliferous with brachiopods dominant. In this district, flints are confined to the upper half of the succession although, elsewhere in Sussex, they are known to occur sparsely down to within a few metres of the base of the formation. This formation is between 40 and 45 m thick in this district and its upper part is clearly exposed with the overlying Lewes Nodular Chalk in the M4 Twyford Down Cutting [490 278] (Plate 1). Here the strata is seen to dip towards the south at about 4°.

### Lewes Nodular Chalk Formation (LeCk)

The Lewes Chalk comprises interbedded, hard to very hard, nodular chalks, with soft to medium-hard chalks and marls. The first persistent seams of flint occur near the base. The flints are typically black or bluish black with a thick white cortex. The formation is generally between 50 and 65 m thick. The Lewes Chalk is divided into two informal units by the paired Lewes Marls and Lewes Flints, comprising a ramifying system of black cylindrical burrow-form flints. The lower unit consists of medium- to high-density chalk and conspicuous, iron-stained, hard, nodular chalks. The upper unit is mainly of low- to medium-density chalks with evenly spaced, thin, hard, nodular beds. The entire Lewes Nodular Chalk Formation can be seen in the Twyford Down M3 motorway cutting (Hopson, 2001a).

Seaford Chalk Formation (SCk)

The Seaford Chalk is composed mainly of soft white chalk with seams of large nodular and semitabular flint. Near the base, thin harder nodular chalks also occur, associated with seams of carious flints, giving this formation a similar appearance to the upper part of the Lewes Chalk. Therefore, the boundary is not clear-cut in mapping terms. Typical brash from the lower part of the Seaford Chalk contains an abundance of fragments of the bivalves *Volviceramus* and *Platyceramus* (Mortimore, 1986a). In the absence of these bivalves, the flaggy bedded nature and pure white colour of the soft chalk serve to distinguish it from the Lewes Chalk below. The Seaford Chalk is 40 to 65 m thick.

Higher in the sequence, the flints are black and bluish black, mottled grey with a thin white cortex, and they commonly contain shell fragments. Towards the very top of the unit (about 5 to 10 m below the Newhaven Chalk), a thin horizon of intensely hard porcellanous silicified chalk, the '**Stockbridge Rock Member**', is commonly developed. It contains abundant sponge spicules, most commonly as moulds, together with some complete sponges and rare echinoids. This lithology is readily identifiable in the brash (forming equant, blocky fragments up to about 5 cm across) and forms a useful marker horizon.

In many places it is associated with a positive topographical feature. It has not been observed in any exposed bedrock section during recent surveys except for a poor section in a silage pit at Beech Farm, Nether Wallop [2845 3537] (Farrant, 2001). The thickness is not known. Field evidence suggests that there may be several thin, hard bands between 5 and 10 m below the base of the Newhaven Chalk, each separated by white chalk. In this district the Stockbridge Rock Member appears widely between Salisbury and Winchester but can be quite sporadic in occurrence. It dies out along the line of the King's Somborne Syncline and becomes patchy farther east into the Alresford district and east of the River Itchen. Its patchy distribution may be explained by variations in the level of cultivation which exposes the characteristic field brash, but it is also likely to be due to lateral changes in the degree of silicification. It occurs at about the level of the Barrois' Sponge Bed and the Clandon Hardground of the North Downs (Robinson, 1986) and may equate with the Whitway Rock of the Newbury area (Sumbler, 1996). In Kent, Rowe's Echinoid Band, a bed of about 30 cm containing an acme occurrence of *Conulus* sp. with other echinoids, occurs just above Barrois' Sponge Bed and is inferred to occur just above the Stockbridge Rock Member.

Newhaven Chalk Formation (NCk)

This formation is composed of soft to medium-hard, smooth white chalk with numerous marl seams and flint bands. Typically, the marls vary between 20 and 70 mm thick (Plate 2). They are much attenuated or absent locally, over positive synsedimentary features, where it is difficult to distinguish between the Seaford and Newhaven formations. Channels with hardgrounds and phosphatic chalks have been recorded elsewhere within the formation (Hopson, 1994; Mortimore, 1986b).

In this district the Newhaven Chalk is estimated to be 40 to 70 m thick. The field brash is composed of smooth, angular, flaggy fragments of white chalk similar in appearance to that of Seaford Chalk. The appearance of abundant flints with *Zoophycos* (a spiral trace fossil usually preserved in flint) near the base of the formation serves as a useful marker for mapping the lower boundary. Individual thecal plates of the zonal index, *Marsupites testudinarius*, occur in numerous small pits, track exposures and as brash, but otherwise macrofossils are rare in the lower part. Just to the west of this district, near East Grimstead [4227 2710], a disused quarry exposes about 40 m of Newhaven Chalk (Mortimore, 1986a), and also good examples of the typical flint and marl bands within the soft Newhaven Chalk. Fossil evidence suggests *Offaster pilula* Zone and *Hagenowia* horizon (Woods, 1999). The uppermost part of the Newhaven Chalk is also exposed in the King's Somborne Quarry (see below).

Culver Chalk Formation (CCk)

The Culver Chalk is up to 70 m thick, and includes the **Tarrant Chalk Member** and the overlying **Spetisbury Chalk Member**, each of approximately the same thickness. The Tarrant Chalk comprises soft, white chalk without significant marl seams, but with some very strongly developed nodular and semitabular flints. The Spetisbury Chalk Member consists of firm, white chalk with large flints, including tabular, paramoudra and potstone forms, with the belemnite *Gonioteuthis* and distinctive forms of the echinoid *Echinocorys*. Some parts of the Culver Chalk (within the *Applinocrinus cretaceus* Subzone) are characterised by abundant bioclastic debris, especially bryozoan debris. The base of the Culver Chalk is taken just below a strong persistent positive feature coinciding with the appearance of abundant large, tabular flint nodules.

On the steeply dipping northern limb of the Dean Hill Anticline, Culver Chalk outcrops within small outliers, many terminating against the Palaeogene deposits of the Alderbury–Mottisfont Syncline. The best exposure is found in a series of small pits near the town of Whiteparish in the southwest of the district. At least 20 m of soft, blocky massive chalk with prominent bedding surfaces and large nodular flints, dipping 3° to the south is exposed. Fossil evidence (Woods, 1999) collected from the pit suggests that much of the upper part of the Culver Chalk is attenuated or cut by channelling, similar to that described by Evans and Hopson (2000) in the Bournemouth area.

Brydone's locality 1067 is described as 'the more southern of two large quarries three-quarters of a mile north of Mottisfont Station'. This is thought to be a very degraded and overgrown pit [3388 2757], facing north, over a minor tributary valley. The poorly located fauna from the recent survey (Woods, 1998a) indicates the Tarrant Chalk Member of the lower Culver Chalk Formation. Farther south and connected to the north-facing outcrop noted above is a much larger and working quarry (King's Somborne Lime Quarry) [3380 2740]. This was visited and comprehensively logged during the survey and up to 49 m of section was recorded (Plate 3). This is, in all likelihood, Brydone's locality 1067 which has been somewhat expanded since his visits. The extensive Brydone collection (regrettably not related to horizons) and that collected during this survey (from marked beds) indicate that the lower part of the exposure, with regularly spaced marls, sponge beds and little flint, is representative of the upper 'belt' of *Offaster pilula*, and hence of uppermost Newhaven Chalk Formation. However, the greater part of the section, characterised by regularly spaced continuous nodular flint beds in soft blocky chalk, is representative of the lower *G. quadratus* Zone, *Hagenowia* Horizon, *A. cretaceus* Subzone (Hopson, 2001b; Woods, 1998, 1999).

Portsdown Chalk Formation (PCk)

The Portsdown Chalk consists of relatively soft, white chalk with common marl seams and some flints; in its lower part there are several horizons rich in inoceramid shell debris. The base is taken at the Portsdown Marl. The top is controlled by the level of erosion below the sub-Palaeogene surface; this surface can cut down to at least the level of the Newhaven Chalk in this district.

The Portsdown Formation only outcrops close to the axis of the Dean Hill Anticline near West Grimstead in the extreme southwest of the district, Whiteparish and possibly near West Dean. There is an estimated 5 m of strata preserved. Only a thin sequence is seen at outcrop (just outside the area of this map) and is exposed in a small, disused quarry near West Grimstead [2172 2648]. The zonation of this exposure is not conclusive but the occurrence of the echinoid *Echinocorys*, belemnites and oysters suggest that the uppermost layers are of Portsdown Chalk age.

## PALAEOGENE

The Palaeogene strata are preserved as a major elongate outcrop within the asymmetrical Alderbury–Mottisfont Syncline and on the dip slopes south of the Winchester Anticline forming the northern margin of the Hampshire Basin. The sequence consists predominantly of clay, silt and sand. Much of the early and late Palaeocene is not represented by strata in this district because at that time the region formed part of a land area separating the Paris and North Sea basins and erosional forces were prevalent.

In latest Thanetian times, deposition occurred in a warm, swampy lowland traversed by braided rivers in which sands, pebble beds and mottled clays of the Lambeth Group were deposited. After a short hiatus, a marine transgression spread from the north, and the district then lay within a broad embayment which included the London, Hampshire, Belgium and Paris basins. The presence of nummulitids attests to a marine connection to the west into the Tethyan Province at this time. The London Clay and Wittering formations were deposited in this broad sea. A hiatus separates the late Ypresian Wittering Formation from the overlying formations of the Bracklesham Group of Lutetian age, reflecting a eustatic sea-level lowstand.

The Palaeogene sequence consists of a number of sedimentary cycles each of which commenced with a marine transgression; the transgressive surface is commonly marked by a thin bed of flint pebbles. These pebbles are overlain by sediments which were deposited as the coastline advanced (Edwards and Freshney, 1987). Each cycle probably lasted between 1 and 2 million years. The structure in this area is dominated by a gentle

**Plate 3** King's Somborne Lime Quarry. View looking south over the working quarry. Newhaven and Culver Chalk Formation [SU 3380 2740] (GS 1096).

southward dip. This general pattern is complicated in places by folding, for example the Alderbury–Mottisfont Syncline which creates an inlier of London Clay to the south of this district.

## Lambeth Group (LMB)

The Lambeth Group corresponds to the strata formerly described as the Woolwich and Reading Beds. Two formations are recognised in this district, the Upnor and Reading formations (Ellison et al., 1994). In mapping it has proved impractical to delineate these separately because the former is very thin and is therefore described here as the basal bed of the Reading Formation.

Reading Formation (Rea)

The Reading Formation rests unconformably on the eroded surface of the Chalk, and is between 20 and 25 m thick. The basement bed of the Reading Formation (equivalent to the Upnor Formation) comprises a reddish brown sand or interbedded sand and clay with abundant rounded to well rounded stained flint pebbles with locally glauconitic sandy clays, analogous to the 'Bottom Bed' of the London Basin. This basal bed is usually less than 1 m thick, at maximum up to 2 m in places. Marked lateral variations in lithology occur along outcrop towards the River Test where sand predominates (Farrant, 1999, Hopson, 2001b).

The Reading Formation consists of mottled, bright red and grey clay and silty clay, but also in shades of purple, brown and orange. The complex mottling has been ascribed to pedogenic processes with multiple overprinting of palaeosols (Buurman, 1980). Lenticular bodies of well sorted, fine- to medium-grained sand occur locally at various levels, particularly at the top and base. Examples of these sands can be clearly seen in Kimbridge Sand and Gravel Pit [SU 323 253] (Plate 4) where finely bedded cross-sets of quartz sand are interbedded with thin silt and clay beds.

In this district, the Reading Formation outcrops in the south-west around the eastward plunging Dean Hill anticline to form the southern limb of the Alderbury–Mottisfont syncline. London Clay masks the Reading Formation in the axis of the syncline, but it reappears on the northern limb around East Tytherley and West Tytherley and eastwards through Braishfield. No clear sections were recorded in this district.

Around Braishfield and Mottisfont, the formation forms the steep face of the escarpment and thin extensions and outliers from it. The basal bed is present throughout the area (it can be augered

**Plate 4** Kimbridge Sand and Gravel Quarry [SU 3230 2530], Reading Formation sands. Finely bedded cross-sets of fine- to medium-grained quartz sand with thinner intervening silt and clay beds. Face is about 2 m high with sandmartin burrows (GS 1081).

wherever ground conditions permit) although it was nowhere exposed at the time of this survey and is usually buried beneath Quaternary deposits at the foot of the escarpment. It is also intimately involved with the development of the clay-with-flints (see p.20). In the east of the district the formation above the basal bed comprises principally red-mottled clay, including variable proportions of silt, and thick sequences of sand and well rounded flint gravel (Plate 5). The colour mottling in shades of red, reddish brown, greyish purple, olive-brown and greenish grey is the result of the oxidation of iron minerals. Where nonoxidised the strata are predominantly grey.

To the west around Timsbury, Michelmersh and Awbridge, the strata comprise thick cross-bedded sand units with lenticular bodies of well rounded 'chatter-marked' flints and only thin mottled clays.

The red mottled clays in this district give rise to heavy wet clay soils that are mainly under pasture and hence there are few natural exposures. The Reading Formation is worked for sand, and to a lesser extent for gravel in the Michelmersh and Braishfield areas. There are no details for the pits in the Michelmersh area, most of which are now degraded. The formation was exposed during the survey in a working sand and gravel pit at Kimbridge, at two disused quarries and at a large gas-transfer development site around Carter's Clay and Kimbridge (see below).

In a large partially landscaped and flooded pit near Carter's Clay, a section [3043 2470] exposed fine- to medium-grained, cross-bedded, quartz sand containing some coarse quartz and angular flint sand, with rip-up clasts, balls and thin lenses of pale grey waxy clay. The sands are bleached white in their upper part (exhibiting a 'pepper and salt' texture in places) becoming yellow-brown with depth and are crosscut by iron-stained (and poorly cemented) oxidation fronts running parallel to the overlying drift deposits. These sands are preserved beneath a variable thickness of head comprising

**Plate 5** Kimbridge Sand and Gravel Quarry [SU 3230 2530]. Exposure of the Reading Formation showing coloured mottled clay resting on a well-rounded flint pebble bed (hammer 30 cm) (GS 1073).

earthy textured silty, clayey sand with pebbles; the base of the head is uneven.

To the north of the exposure described above a large site was being developed as a gas-transfer station. In the extensive exposures, a variety of facies within the Reading Formation were visible, principally on a cleared inclined surface with minor small 'cuts', totalling some 6 m of beds. Elsewhere at the site, clay containing well rounded flints grades up into a soil.

Michelmersh Brickworks [343 258] still produces bricks for the specialist market from the Reading Formation. There are a number of pits in the area but it is not known whether they all still deliver clay to the main works. White (1971) claimed that the brickworks utilises clay from the London Clay and Reading Formations but all of the local pits within the Michelmersh area exploit the Reading Formation. White (1971) gave the following account of the working method, but it is not known whether the bricks are still hand moulded. 'The clay is brought to the pit by a Crawler tractor to the feeder and then the pug mill. Here breeze is added by hand before being mixed and passed on to the brickmakers' benches. The green bricks are set on pallets and transferred to the drying sheds which utilise the excess heat from the kilns. They are left for a week before being set in a kiln for burning. The pair of kilns are dome shaped, circular down-draught kilns, hold between 35 and 38 thousand bricks and are 20 feet in diameter. They are oil fired and reach a temperature of 1050 degrees Centigrade and it is this heat which produces the yellow, brown, deep red, purple and blue colours for which Michelmersh bricks are known'.

## Thames Group

London Clay Formation (LC)

This formation, within the Thames Group, consists mainly of grey, pyritic, bioturbated, silty and fine-grained sandy clay with interbedded seams of calcareous cementstone and rounded flint pebble beds; a glauconitic sandy bed occurs at the base ('Basement Bed'). The formation contains sheet-like and lenticular bodies of fine-grained sand which generally mark the top of coarsening-upward sedimentary rhythms (King, 1981). Each rhythm, when complete, has a basal pebble bed of a richly glauconitic horizon that passes up into silty clay, which becomes progressively more silty and sandy upwards. The rhythm is completed by cross-bedded sand or interbedded channel-fills. Some of these sands may be very shelly such as the informally named 'Lingula Sands' (Meyer, 1871). The lithological changes in each rhythm reflect an early marine transgression, followed by low-energy marine sedimentation and a final progradation of coarse sediment from the margins of the depositional basin. The thickness of the formation varies greatly; between 30 and 85 m, with the greatest thickness in this district beneath the axis of the Alderbury–Mottisfont Syncline. The London Clay Formation was formerly worked for brick clay near East Tytherley [2910 2870] and West Dean [2600 2830] where there are several poor sections of orange-brown, silty, pebbly, micaceous clay (Farrant, 2001).

There are no natural exposures of this formation in the district although drainage ditches show up to 1.5 m of pale yellow-brown, silty, fine-sandy clay grading down into pale and dark grey mottled silty clay in the deeper or freshly cleaned ditches.

## Bracklesham Group

This group contains a varied succession of interbedded clay, silty and sandy clay, silt and sand. Shell, lignite and pebble-bed horizons throughout the succession reflect deposition in transgressive/regressive sedimentary cycles. The group is divided into four formations, of which only the lower Wittering Formation is preserved in this district.

Wittering Formation (Wtt)

The Wittering Formation occurs as small outliers around Awbridge [320 243] in the extreme south of the district. There are three main lithologies thought to have been deposited in environments ranging from intertidal mud and sand flats to the subtidal zone. The first and most widespread in this district is clay dominated, comprising olive-grey clay and sandy clayey silt with lenses of very fine-grained sand. The second comprises sand and silty sand with lignite and pyritised bivalves interbedded with clay. The third consists of fine- to medium-grained, sparsely glauconitic sand with silty clay laminae. The three main lithologies interfinger both laterally and vertically; it was not possible to map them individually due to the complex interdigitation of lithologies and paucity of exposure.

The base of the Wittering Formation is taken at the base of thinly bedded to laminated clays or sands that rest on a variety of London Clay lithologies, most noticeably the fine- to medium-grained sands of the Whitecliffe Sand Member in this area. The formation overall is up to 35 m thick in this district. There are few exposures and the numerous pits described by Reid (1903) are no longer visible. There are no natural exposures of these soft, easily degraded deposits and the outcrop is delimited by augering.

## QUATERNARY

About 60 million years is estimated to have elapsed between the deposition of the youngest preserved Palaeogene and the oldest Quaternary deposits in this district. During this time younger Palaeogene and Neogene strata were deposited across much of southern Britain, and subsequently were removed following uplift along the Wealden axis (as part of the general inversion of the Wessex Basin). During the Quaternary, a further significant break in deposition occurred after the initial accumulation of the clay-with-flints and before the deposition of the younger Pleistocene deposits.

During the Pleistocene, sea levels rose and fell according to the quantity of water locked up in ice caps. At times of glacial maxima, a periglacial environment was established in this district. There was enhanced erosion both by solifluction and by an extensive river system flowing to much lower base levels (up to 100 m below present sea level in the most extreme glacial episodes).

The following descriptions of the deposits are grouped on the basis of their origin. Mass movement deposits are described first, followed by fluviatile deposits. Their order does not imply relative age.

### Clay-with-flints

Clay-with-flints is typically composed of orange-brown or reddish brown clays and sandy clays containing abundant flint nodules and rounded pebbles. At the base of the deposit the matrix is stiff, waxy and fissured (slickensided), and dark brown in colour. Relatively fresh nodular flints are stained black and/or dark green, possibly by manganese compounds and/or glauconite. The deposit gives rise to a stiff, red-brown, silty clay soil, strewn with flints. This is primarily a *remanié* deposit resulting from the modification of the original Palaeogene cover and dissolution of the underlying chalk (Plate 7). The thickness of the clay-with-flints is about 5 to 6 m, as a general maximum, but this may rise to over 10 m in limited areas, usually where dissolution of chalk is most pronounced. The margin of the clay-with-flints is sharply defined on the scarp edge but the boundary is diffuse on the chalk dip slope. This downslope feather edge is obscured by a lateral passage into a late-stage solifluction deposit or head gravel, distinguished with the prefix 'G' on the map. These deposits have a more sandy matrix and a surface brash composed principally of gravel-sized broken angular flints.

### Head

In general, head comprises yellow-brown, silty, sandy clay with variable proportions of coarser granular material, but all deposits have an earthy texture. Clast composition varies depending on source materials; those deposits derived mainly from the chalk were formerly mapped as 'dry valley deposits' or 'coombe deposits'. These heterogeneous deposits accumulated in valley bottoms by solifluction, hillwash and hillcreep and are generally only a few metres thick. Head deposits occur as soliflucted slope deposits and gravelly valley bottom deposits. The slope deposits are thought to represent an earlier phase of deposition before the valley bottom head.

### Slope Head

Slope Head ('Older Head') deposits range from flinty gravels to reddish brown, sandy clays containing abundant flint nodules and pebbles that are generally much more shattered than those in the clay-with-flints. Several large sheets occur in this district, generally no more than a few metres thick. The deposits are most widespread on north- and east-facing slopes and commonly grade laterally into areas with only a thin flinty veneer.

Gravelly Head

These broad, sheet-like deposits are composed mainly of angular flint gravel set in a stiff, sandy clay matrix. In some exposures the matrix contains chalk, but elsewhere this has been lost by decalcification. The head gravel is generally regarded as the result of solifluction of chalk, Palaeogene deposits and clay-with-flints down the dip slope of the Chalk during cold phases of the Quaternary.

River Terrace Deposits

River Terrace Deposits are associated with the major river systems of the Test and the Itchen. The River Test flows north to south across the centre of this district and its catchment covers the greater part of Sheet 299 Winchester. Terrace aggradations are found throughout the district. They are best developed in terms of lateral extent and vertical height range near to the confluences of the rivers Dever and Anton in the north near Wherwell, and in the south with the River Dun near Mottisfont. Up to eight aggradations have been mapped.

The terrace deposits consist generally of gravel and sandy gravel. Clasts comprise subangular to angular flint with some well rounded and nodular flint, and subordinate quartz, ironstone, sarson and other exotic debris (Hopson, 2001b). Some clasts are rubified, probably derived from the clay-with-flints. Gravels are both matrix- and clast-supported, and planar to cross-bedded. The sand matrix is predominantly medium-grained quartz with some fine- and coarse-grained quartz; the coarser fraction contains ironstone and flint shards. The lower terrace aggradations are generally 'clean' with little silt and clay content whereas higher terraces become progressively more 'clayey'. Iron cementation has occurred in places (Hopson, 2001b).

The River Itchen flows across the south-east of the district, through the city of Winchester and southwards past Twyford. There are three terrace aggradations, with the deposits being predominantly clayey sandy gravels. They are mapped principally on the western side of the valley indicating that the stream migrated eastwards. However, this is contradictory to the south-westward dip of the strata and may indicate a structure (e.g. fault) buried beneath the broad alluvial tract (Hopson, 2001a). In places clayey and sandy silt and silty clay mask the underlying aggregate, perhaps indicating preservation of overbank or aeolian deposits at the top of each fluvial cycle. Flint again predominates, together with chert, polished fine-grained quartz, and some clasts are rubified. In general the terraces are up to 5 m thick.

There is little direct evidence of the age of the terraces, but most aggradations are probably periglacial in origin. Terrace deposits above the second terrace show cryoturbation structures indicating that they have suffered at least one periglacial event, and thus suggesting they are all pre-Devensian.

Alluvium

The alluvium comprises soft, organic, mottled silty and sandy clay which generally overlies a basal lag gravel. Thin stringers of gravel may occur within the sequence, indicating channel migration or periodic increases in the flow regime of the rivers over time. In general the deposit is thin, usually between 1 and 3 m in the upper reaches of rivers, but at major confluences and in the lower reaches of the rivers up to 8 m have been proved. A common characteristic of streams flowing over chalk bedrock is the association of calcareous tufa and peat accumulations with the overbank alluvium. Major occurrences of these deposits are discussed below. Plate 6 shows a typical bankside exposure of the floodplain of the River Test near Kimbridge [3320 2630]. Grey silty clay overbank deposits are closely associated with variably silty and clayey calcareous tufa both resting on a basal bed of dark fibrous peat.

## Tufa

Extensive patches of tufa occur in the Test and Itchen valleys. The tufa commonly forms raised (1 to 2 m) hummocky spreads, superficially very similar to river terraces on the valley floor and for the most part, the distribution shown on the map reflects the occurrence of this characteristic morphology. The deposit usually consists of small amorphous grains or nodules of calcium carbonate, commonly coating flint pebbles or other nuclei. Mollusc and other shell fragments are locally common. The level of the raised hummocks probably represents the winter flood level of the water or marsh before the river channels were straightened and the marshes drained.

## Peat

Small accumulations of peat and peaty material are associated with alluvium and the river terrace deposits but these are generally too limited in extent to map. More extensive spreads of peat were mapped in the Kimbridge to Timsbury area where it consists principally of fibrous wood and sedge material. Commonly the peat is seen to pass under, or grade into, alluvium and calcareous tufa. Elsewhere peat is noted beneath overbank silty clay deposits along many of the drainage channels. In all occurrences there is an intimate association with calcareous tufa and there is a continuum between fibrous pure peat and silty clay with a high humic content (both finely disseminated and as lenses of limited thickness and extent).

## Artificial deposits and Worked Ground

The more extensive areas of artificial deposits and worked ground are shown on the map but many minor occurrences have been omitted for clarity. The 1:10 000 scale maps of the district, listed in *Information sources* show the distribution in more detail. Artificial ground is common, especially within the urban conurbation and associated with the development of major routeways.

## Made Ground

Made ground is shown in areas where material was deposited by man upon the natural ground surface. There are two main categories:

- natural materials produced either as spoil from mineral extraction, or dug for

**Plate 6** Typical bankside exposure on the floodplain of the River Test near Kimbridge [SU 3320 2630]. Layers of humic silty sandy clay overlying granular calcareous tufa and a basal fibrous dark peat (GS 1116).

the construction of various embankments and raised areas, including bunds for flood defence
- waste in landfill sites; recycling of waste construction materials is leading to their increased usage in urban and industrial landscaping

### Worked Ground

Worked ground is shown where natural materials are known to have been removed, for example in quarries and pits, road and rail cuttings and general landscaping. There are a number of disused pits and quarries in this district associated with the extraction of Chalk, terrace gravels, London Clay and Reading formations. Most lie along, or close to, the major rivers.

Near Mottisfont on the River Test is a large working quarry (Somborne Lime Quarry) [3380 2740] where the Newhaven and Culver Chalk Formations are extracted. Another large partially landscaped and flooded pit is situated near Carter's Clay [3043 2470]. An extensive area of old and current workings occurs between Dunbridge Hill and Kimbridge where extraction of sand and gravel has taken place from the second to third terrace sequence of the Test and the underlying Reading Formation. Kimbridge Quarry [3215 2550] is presently worked (Reading Formation) as mentioned above, and Michelmersh Brickwork's [343 258] still produces bricks for the specialist market from the Reading Formation.

### Infilled Ground

Infilled ground comprises areas where the natural ground has been removed and the void, wholly or partly backfilled with man-made deposits which may be either natural or waste material, or a combination of both. Where quarries and pits have been filled, the ground restored and landscaped, built on or returned to agricultural use, there may be no surface indication of the extent of the backfilled area. In such cases the boundaries of these sites is taken from archival sources such as earlier aerial photographs, local authority records and old topographical and geological maps.

### Landscaped Ground

This consists of areas that have been extensively remodelled or landscaped, with complex patterns of cut and fill, too small to be identified separately. Such areas commonly include parkland, golf courses and major construction sites.

## STRUCTURE

The district lies on the broad chalk plain between the Wealden Basin (anticline) in the east and the Wessex Basin to the west. The Wessex Basin is situated between complex folded and faulted structures associated with, to the south, the Wardour-Portsdown Inversion and, to the north, the Pewsey Anticline around Burbage and the Vale of Pewsey (Chadwick, 1986, 1993). The southern structure is expressed as the east–west-trending Mere Fault (Figure 10 GL2) farther to the west through the Vale of Wardour and is imaged at depth in seismic and gravity data.

The main structures affecting the Weald are thought to have formed during the Miocene. The resultant structure in the western Weald is a broad, gentle anticline with shallow dips (less than 4°) to the south, variable dips (up to 20°) to the west or north-west, and a gentle plunge west-south-westwards. Periclinal structures associated with east–west faulting at depth are known to be associated with the Wealden Anticline. Within this broad structure are smaller, lower amplitude synclines and anticlines. In general, the small anticlinal structures have steeper dips on their northern limbs, probably reflecting the reactivation of deep-seated Jurassic fault blocks.

There is growing evidence that the Late Cretaceous was not a tectonically quiescent period, but that continued tectonic and

# GEOLOGICAL DESCRIPTION

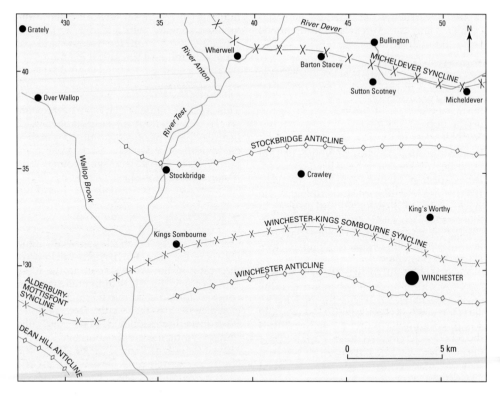

**Figure 7** Structure of the district showing the approximate position of the axial traces of principal folds.

eustatic movement occurred in phases throughout (Mortimore et al., 1998). Four major tectonic phases (demonstrated in Germany and the eastern Anglo–Paris Basin) generated channelling, slumping, hardground and various flint development, phosphatic chalks as well as variations in marl development throughout Southern England. Certain characteristics of the chalk in this district, such as the very hard, silicified horizon, the Stockbridge Rock Member (Seaford Chalk Formation), may have been a result of this continued movement. Features within the chalk could therefore be a result of, or effected by, both syndepositional and postdepositional tectonics.

Faults within the Jurassic strata at depth control the location of many of these fold structures. One such fold pair is the Winchester Anticline and its associated Winchester–King's Somborne Syncline to the north (Figure 7). This can be traced eastwards to the Alresford district as the Winchester–Meon structure (pericline) which plunges gently to the west, mirroring the axis of the Wealden Anticline. The axis of a similar but broader anticline, the Stockbridge Anticline, runs through the centre of the district but gradually dies westwards. To the north, around the Wallops and Stockbridge, the dip is less than 0.5° to the south-west; to the south of the axis, the dip is 1° to 2° to the south. Its counterpart, the Micheldever Syncline farther north, trends east–west before veering northwards towards Andover.

Another major structure in this district is the Dean Hill Anticline and associated Alderbury–Mottisfont Syncline to the north. The Dean Hill Anticline is similar in orientation and form to the Winchester Anticline east of the Test valley. Both these folds are periclinal structures with east–west-trending axes. The Dean Hill Anticline is an asymmetric pericline with a shallow 1° to 2° dip to the south and a much steeper 8° to 12° dip on its northern limb. It plunges eastwards at around 1° to 2°. The core of the anticline brings the lower Newhaven Chalk to surface around West Dean (Farrant, 2001); to the north, the core of the corresponding Alderbury–Mottisfont Syncline is occupied by the London Clay Formation. The northern limb of the syncline has a gentle southerly dip around 1° to 2°. Structure contours created on the base of the Stockbridge Rock Member and the base of the Palaeogene highlight the folding and regional structure of the district (see Sheet 299 Winchester).

The periclinal structure centred on Cheesefoot Head (Winchester anticline) was intensely investigated for hydrocarbons. The information from a few deep boreholes into the Jurassic was enhanced by additional shallower boreholes proving the Melbourn Rock and Glauconitic Marl. This structure is known to have dips of 10° to the north of Spitfire Bridge [497 294] on its northern limb, and dips of 3° to 4° to the south-west (Twyford Cutting) on its southern limb.

There are few recognisable faults at surface, although small faults with throws of less than 1 m can often be seen in pits and quarries but cannot be traced farther. However, seismic surveys often indicate faulting at depth (in the Jurassic strata) which coincides with fold axes at surface. Instead of discrete faults, the strata may be displaced by several metres by numerous minor shear zones over a distance of several tens of metres, thus smearing the 'fault' zone over quite a wide area. The faults become attenuated within the softer chalk sequence, and are expressed at surface as broad folds.

The notable exception is the Over Wallop Fault. This extends from just west of Castle Farm [2700 3930] and runs east-north-east through Kentsboro [3050 4035] and on towards Red Rice [3350 4150] where it appears to die out (Farrant, 2001). It cannot be traced westwards because of lack of exposure across the military area of Porton Down. Total displacement of the fault is estimated to be approximately 15 to 20 m, downthrowing to the south. It displaces the Stockbridge Rock Member and the Newhaven/Seaford Chalk boundary. A second, similar but much smaller fault was identified near Stockbridge [3850 3590] along the line of the A30 trunk road. This also displaces the Stockbridge Rock Member and Newhaven/Seaford Chalk boundary. This fault downthrows to the south by approximately 5 to 10 m.

The Lewes Chalk outcrop is limited southward at Tilebridge Farm [344 333] by a fault trending west-north-west and downthrowing to the south. Micropalaeontological evidence (Wilkinson, 2000) demonstrates very basal Seaford Chalk up-slope from the farm and this together with a grainy and nodular chalk and carious flint surface brash are sufficient to justify the boundary shown on the map at about 48 m OD. The fault must have a throw of at least 18 m as the Lewes Nodular Chalk does not reappear to the south of the fault on the river bluff adjacent to the floodplain, which is itself at 30 m OD.

# 3 Applied geology

Geological factors have a bearing on the siting and nature of future urban and industrial developments. By giving consideration to geological conditions at an early stage in the planning process, it may be possible to mitigate some of the problems commonly encountered during construction work. The diverse local geology gives rise to variable ground conditions and some significant aspects are discussed below. Other important factors are water resources, mineral workings and the risk of flooding.

## Hydrogeology

The Chalk is the major aquifer in the district and has the largest storage capacity and catchment area, providing the major supply of potable water. The chalk aquifer is confined in the south of the district by the Palaeogene cover. In the unconfined aquifer, the water table broadly follows the surface topography in subdued form, with subsurface flows away from recharge areas on high ground. Natural annual fluctuations in the unconfined aquifer can exceed several metres under the high parts of the downs. In the confined aquifer, natural fluctuations of the potentiometric surface decrease away from the outcrop. The aquifer also contributes to the baseflow of the rivers draining southwards across the district.

The sandy beds of the Lower Greensand Group constitutes a separate aquifer beneath the Gault aquiclude. Water is also obtained in small quantities from the Palaeogene and to a lesser extent, the Quaternary, but supplies are variable in both quantity and quality.

Yields from the Chalk Group of the Winchester district (Hargreaves, 1981) vary between the formations typically producing 10.5 l/s (litres per second) from boreholes in the traditionally named 'Upper Chalk' (Lewes to Portsdown Formations), to 2.3 l/s in the lower, marly chalks (West Melbury Chalk Formation). The highest yields, recorded elsewhere, such as the Fareham area, show that up to 270 l/s are obtained from large-diameter shafts, boreholes and headings in the White Chalk Subgroup.

The hydraulic properties of the Chalk aquifer are complex and result from a combination of matrix and fracture properties. The Chalk is microporous with low intrinsic permeability. The intergranular porosity of the Chalk is high, usually around 35 per cent for the White Chalk Subgroup, falling to around 25 per cent for the Grey Chalk Subgroup (Bloomfield et al., 1995). However, the pore sizes are so small that the permeability of the rock is minimal. The high transmissivity of the aquifer is provided by fractures, which are commonly enlarged by dissolution.

Rapid groundwater flows are sometimes found in the unconfined Chalk aquifer where karstic-type development has taken place. This is commonly associated with the proximity of thin cover, such as the Palaeogene deposits or clay-with-flints. For example, tracer studies on the floodplain of the River Pang in Berkshire have revealed rapid flows (velocities of over 6 km/d) between swallow holes and a spring known as the Blue Pool (Banks et al., 1995).

Chalk water is usually of very good chemical quality. At outcrop it is hard to very hard, with carbonate hardness predominating. However, at depth in the confined aquifer, the water changes to a soft sodium bicarbonate type as a result of ion exchange. At the same time chloride ion and total dissolved solids increase (though not beyond potable limits), fluoride increases and anaerobic conditions set in, where nitrate is replaced by ammonia (Institute of Geological Sciences, 1978).

The Upper and Lower Greensand and the Palaeogene sandstones and sands are porous, essentially nonfissured aquifers, although the Upper Greensand is loosely indurated with some fissuring.

Wells in the chalk are generally unlined, while those in the sands require screening. Perennial springs occur at the junction between the Chalk and overlying Palaeogene. A number of these are recorded around the West Dean area. Valleys on chalk bedrock are normally devoid of surface water, although during extreme weather conditions, torrential rain over successive days can result in surface flow and groundwater flooding.

## Bulk minerals

### Chalk

There are many pits in the district attesting to the great historical use of chalk. The widespread digging of chalk for marling adjacent to loamy land and acidic soils, for burning to produce agricultural lime and as a source of flint for building goes back at least to Roman times. Most of these pits are abandoned and degraded, some have been utilised for disposal of inert domestic and industrial waste, but agricultural lime may still be produced from some of the smaller privately owned pits for use on individual farms. The only large quarrying operation at the time of survey, the Somborne Lime Works, produces agricultural lime and, as a by-product, flints for building restoration.

### Sand and gravel

Throughout the district sand and gravel has been worked from the river terrace deposits that occupy the interfluves, valley sides and floors. Sand has also been excavated sporadically from the Palaeogene Reading Beds and Wittering Formations. Large volumes have been taken from the Reading Formation at the Kimbridge Sand and Gravel Pit [SU 323 253] just south of Mottisfont on the western bank of the River Test. This quarry is still in operation (2001).

### Clay

Local sources of brick clay include the London Clay, 'plastic clays' within the Bracklesham Group and the mottled clays of the Reading Formation. Brick manufacture has been an important industry in the past with clays, mainly from the Reading Formation being extracted. Today, Michelmersh Brickworks [343 258] still produces bricks for the specialist market. There are a number of pits in the area but it is not known whether they all still deliver clay to the main works.

### Building stone

Building stone is not produced commercially in this district, but in the past locally derived materials have been used in construction. Limited use has been made of the harder beds within the Chalk sequence (Melbourn Rock Member, Lewes Nodular Chalk Formation). Flint, as a 'waste' product of chalk extraction and from 'field picking', has also been used to maintain farm tracks and as a source of decorative 'dressed' flints for buildings.

## Hydrocarbons

The district was first explored for hydrocarbons in the 1930s and, more successfully, in the 1980s with the discovery of the Humbly Grove Oil Field just to the north-east of the district. Other oil fields have been found in the Weald Basin at Horndean, Singleton, Storrington and Stockbridge within this district. The reservoir rocks are in the Jurassic Great Oolite Formation.

The process of hydrocarbon formation, migration and entrapment is controlled by east–west, pre-Albian extensional faults. The Humbly Grove oilfield, as an example, is developed on a clearly defined, tilted horst block bounded by two such (now reversed) extensional faults. In the basins to the south of the major faults, and particularly in the centre of the Weald Basin, the Lias Group source rocks, and possibly the Oxford Clay and Kimmeridge Clay were buried sufficiently deep to generate hydrocarbons. The hydrocarbons migrated south from the centre of the Weald Basin into the Great Oolite rocks of the palaeo-highs, where antithetic faults provide traps. Migration probably began in Early Cretaceous times and may have continued until uplift in mid-Tertiary times (Penn et al., 1987). Although Cainozoic compression caused inversion of the Weald Basin, it did not destroy all the traps. Many anticlines in both the Weald and Wessex basins, for example the Portsdown and Littlehampton anticlines, do not contain oil and gas, suggesting that primary oil migration ceased before they were formed.

### Geotechnical considerations

There is a range of potential problems relevant to ground stability in the district. Figure 8 tabulates potential ground constraints and the deposits with which they are commonly associated. The following statements should be taken only as a guide to likely or possible problems and should not replace site-specific studies.

The relatively loose sands of some of the units within the Palaeogene strata provide unreliable foundations on steep slopes. Freshly ploughed fields or exposed ground can become gullied during heavy rainfall.

The London Clay Formation contains clays with a high smectite content that is susceptible to shrinking and swelling. Consequently they may swell on wetting or crack during extreme drought conditions. Suitable precautions should be taken during construction.

Peat, and other alluvial deposits which contain thin beds of peat, may be liable to compression and differential compaction when the ground is subject to loading.

Most natural slopes are thought to be stable in the district but this can be strongly influenced by human activity, particularly where over-steepened slopes on Palaeogene deposits are created during construction work.

### Chalk dissolution

Chalk is prone to dissolution because of the action of acidic rainwater and groundwater, especially in the near surface vadose zone. This process was more enhanced during prolonged periods of periglacial conditions in the Pleistocene. The resultant swallow or sink holes, which can measure up to 50 m in diameter and 6 m deep, are common features of chalk outcrop in southern Britain, but distinguishing them from small, old chalk pits can be difficult.

As a consequence of dissolution, fractures occurring naturally in the chalk are enlarged. The resultant pipes, which may be filled with clay-with-flints (Plate 7), continue to provide sumps for excess surface water, making them liable to further subsidence and differential settlement. Solution features are likely to be common on the outcrop of the younger Chalk formations, particularly where a thin clay-with-flints or Palaeogene cover occurs nearby.

The chalk adjacent to the feature is rubbly in texture indicating preferential dissolution along fractures and joints. Many

**Plate 7** Solution pipe infilled with clay-with-flints; near King's Somborne Quarry [SU 3380 2740] (GS 1159).

**Figure 8** Potential ground constraints.

| Geological unit | Potential ground constraints |
|---|---|
| Worked ground | variable foundation conditions<br>unstable sides on old workings |
| Made ground | variable foundation conditions<br>leachate and methane production from waste |
| Infilled ground | as above |
| Disturbed ground | slope instability<br>variable foundation conditions |
| Head | variable foundation conditions<br>ground heave |
| Peat | compressible strata<br>risk of flooding |
| Alluvium | compressible strata<br>risk of flooding<br>variable foundation conditions |
| River terrace deposits | high water table<br>possibility of undocumented and filled former pits |
| Wittering Formation | local perched water tables<br>ground heave in clay members<br>loose sand prone to erosion and gullying |
| London Clay Formation | ground heave<br>landslip and subsidence in clays<br>high sulphate content groundwater<br>perched water table and springs in sand layers |
| Reading Formation | variable foundation conditions<br>ground heave<br>sink holes close to contact with Chalk<br>perched water tables and springs in sand layers |
| Chalk Group | slightly elevated natural radon emissions<br>groundwater protection requirement<br>possibility of undocumented and infilled former pits<br>dissolution cavities and sinkholes |

borehole logs record a comparable upper rubbly top to the Chalk, particularly beneath the lower river terrace deposits where the Chalk is commonly in contact with flowing groundwater.

Chalk has a high natural water content and this may lead to slurrying if over compacted. The stability of excavations in chalk is largely controlled by the frequency and direction of natural cavities and joints.

In addition to the naturally occurring hazards, man has had considerable influence on the landscape. Many of the abandoned aggregate, chalk and clay pits in the area have been filled, particularly adjacent to urban areas. Records are held by the local authorities, but old areas of fill are often poorly documented. Cuttings and embankments for major road and rail links are commonplace in the district.

# Information sources

Enquiries concerning geological data for the district should be addressed to the Manager, National Geological Records Centre, BGS, Keyworth.

Other geological information held by the British Geological Survey include borehole records, fossils, rock samples, thin sections, hydrological data and photographs. Searches of indexes to some of the collections can be made on the Geoscience Data Index system in BGS libraries and on the web site; a *Catalogue of maps and books* is available on request (see back cover for addresses).

## BGS reports

Technical reports relevant to the district are included in the bibliography. Most are not widely available but may be consulted at BGS and other libraries.

BIOSTRATIGRAPHY  There are nine biostratigraphical reports by the authors M A Woods and I Wilkinson, covering the Winchester district. These are held as internal open file reports by the Biostratigraphy Group of BGS. Readers are recommended to contact the Group Manager, Biostratigraphy, BGS, Keyworth for access to these reports and *Palaeontological collections*.

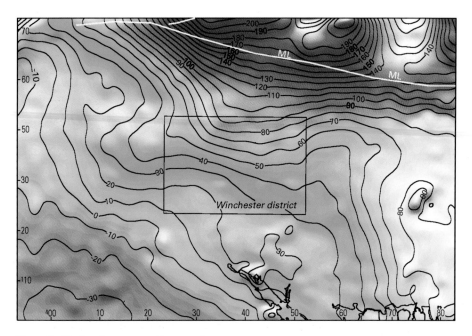

**Figure 9**  Regional aeromagnetic anomaly map.
Total field magnetic anoamlies in nanotesla (nT) relative to a local variant of IGRF90.
Survey flown at a constant barometric height of 549 m on N–S flight lines 2 km apart and E–W tie lines 10 km apart.
The anomalies are shown as a colour shaded relief presentation using the BGS COLMAP Package.
The shaded topographic effect has been created using an imaginary light source, located in the north.
Contour interval 10 nT.
Lineament ML, taken from Busby et al. (2001), marks the Variscan Front and separates high amplitude anomalies in the north from those of longer wavelength and lower amplitude to the south.

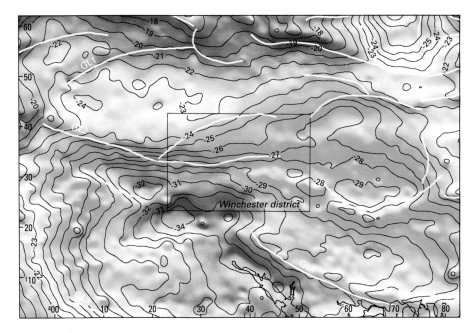

**Figure 10** Regional Bouguer gravity anomaly map.
Bouguer gravity in milligals (mGal) calculated against the Geodetic Reference System 1967, referred to the National Gravity Reference Net 1973. a variable reduction density has been used.
The anomalies are shown as a colour shaded relief presentation using the BGS COLMAP Package. The shaded topographic effect has been created using an imaginary light source, located to the north. Contour interval 1 mGal ($1 \text{ mGal} = 1 \times 10^{-5} \text{ m/s}^2$).
Lineaments shown are taken from Busby and Smith (2001) and correlate with major structures such as the Pewsey (GL1), Mere (GL2), Cranborne (GL3) and Wardour–Portsdown (GL4) faults.

## Maps

For the most recent revision of the 1:50 000 Series Sheet 299 Winchester, the component *1:10 000 series geological maps* are available for purchase from the BGS Sales Desk.

*Geophysical colour shaded relief maps* of gravity and magnetic anomalies for Britain, Ireland and adjacent areas are available at a scale of 1:1 500 000 (Figures 9; 10).

Small scale regional *geochemistry, hydrogeology* and *mineral* maps that include this district are also available from BGS.

## BGS collections

The *Palaeontological collections* held at BGS Keyworth includes macrofossils and micropalaeontological samples collected from the district; enquiries should be directed to the Chief Curator, BGS, Keyworth.

The *Petrological collections* includes hand specimens and thin sections. Further information, including methods of accessing the database, charges and conditions of access to the collection are available on request from The Manager, Petrological Collections, BGS Keyworth.

*Bore core collection*, held at BGS Keyworth, contains samples and entire core from a small number of boreholes in the Winchester district.

Copies of the photographs used in this report (excluding the cover photograph) are included in the *BGS Photographic collection*, and can be supplied at a fixed tariff.

# Selected bibliography

Most of the references listed below are held in the library of the British Geological Survey at Keyworth, Nottingham. Copies of the references may be purchased from the library subject to the current copyright legislation. BGS Technical Reports for this and adjacent areas are included in this bibliography. The reports may be purchased from BGS or consulted at BGS and other libraries.

ALDISS, D T. 2000. Geology of the Cholderton and Grately area, Hampshire and Wiltshire. *British Geological Survey Technical Report,* WA/00/11.

BANKS, D, DAVIES, C, and DAVIES, W. 1995. The Chalk as a karstic aquifer: evidence from a tracer test at Stanford Dingley, Berkshire, UK. *Quarterly Journal of Engineering Geology*, Vol. 28, S31–38.

BLOOMFIELD, J P, BREWERTON, L J, and ALLEN, D J. 1995. Regional trends in matrix porosity and dry density of the Chalk of England. *Quarterly Journal of Engineering Geology*, Vol. 28, S131–142.

BOOTH, K A. 2000. Geology of the Farley–Pitton district, Wiltshire. *British Geological Survey Technical Report,* WA/00/59.

BOOTH, K A. 2001. Geology of the Wherwell–Bullington district, Hampshire. *British Geological Survey Technical Report,* WA/01/08.

BOOTH, K A. 1999. Geology of the Wonston–Leckford district, Hampshire. *British Geological Survey Technical Report,* WA/99/10.

BRISTOW, C R. 1998. Geology of the Micheldever district, Hampshire. *British Geological Survey Technical Report,* WA/98/62.

BRISTOW, C R. 1999. The stratigraphy of the Chalk Group of the Wessex Basin. *British Geological Survey Technical Report,* WA/99/08.

BRISTOW, C R, BARTON, C M, FRESHNEY, E C, WOOD, C J, EVANS, D J, COX, B M, IVIMEY-COOK, H I, and TAYLOR, R T. 1995. Geology of the country around Shaftsbury. *Memoir of the British Geological Survey,* Sheet 313 (England and Wales).

BRISTOW, C R, MORTIMORE, R N, and WOOD, C J. 1997. Lithostratigraphy for mapping the Chalk of southern England. *Proceedings of the Geologists' Association*, Vol. 108, 293–315.

BRYDONE, R M. 1912. The stratigraphy of the Chalk of Hants. (London: Dulau).

BUSBY, J P, and SMITH, N J P. 2001. The nature of the Variscan basement in south-east England: evidence from integrated potential field modelling. *Geological Magazine*, Vol. 138, 669–685.

BUURMAN, P. 1980. Palaeosols in the Reading Beds (Palaeocene) of Alum Bay, Isle of Wight, UK. *Sedimentology*, Vol. 27, 593–606.

CHADWICK, R A. 1986. Extension tectonics in the Wessex Basin, southern England. JOURNAL OF THE GEOLOGICAL SOCIETY OF LONDON, Vol. 143, 465–488.

CHADWICK, R A. 1993. Aspects of basin inversion in southern Britain. JOURNAL OF THE GEOLOGICAL SOCIETY OF LONDON, Vol. 150, 311–322.

EDWARDS, R A, and FRESHNEY, EC. 1987. Geology of the country around Southampton. *Memoir of the British Geological Survey,* Sheet 315 (England and Wales).

ELLISON, R A, KNOX, R W O'B, JOLLEY, D W, and KING, C. 1994. A revision of the lithostratigraphical classification of the early Palaeogene strata of the London Basin and East Anglia. *Proceedings of the Geologists' Association*, Vol. 105, 187–197.

EVANS, D J, and HOPSON, P M. 2000. The seismic expression of synsedimentary channel features within the chalk of southern England. *Proceedings of the Geologists' Association*, Vol. 111(3), 219–230.

FARRANT, A R. 1999. Geology of the King's Somborne–Winchester district, Hampshire. *British Geological Survey Technical Report,* WA/99/06.

FARRANT, A R. 2001. Geology of the Dean and Wallops area, Hampshire and Wiltshire. *British Geological Survey, Technical Report,* WA/00/11.

FARRANT, A R and BOOTH, K A. 2001. Geology of the Danebury area, near Andover, Hampshire. *British Geological Survey Internal Report,* IR/01/139.

HANCOCK, J M. 1975. The petrology of the Chalk. *Proceedings of the Geologists' Association*, Vol. 86, 499–535.

HARGREAVES, R, PARKER, J, and TASKIS, D M. 1981. Records of wells in the Winchester area. *Metric*

# SELECTED BIBLIOGRAPHY

well inventory of the Institiute of Geological Sciences, Sheet 299.

HODGSON, J M, CATT, J A, and WEIR, A H. 1967. The origin and development of clay-with-flints and associated soil horizons on the South Downs. *Journal of Soil Science*, Vol. 18, 85–102.

HOPSON, P M. 1994. Geology of the Treyford, Cocking and Chilgrove district, West Sussex. *British Geological Survey Technical Report*, WA/94/48.

HOPSON, P M. 1998. Geology of the area around New Alresford and Cheriton, Hampshire. *British Geological Survey, Technical Report*, WA/98/50.

HOPSON, P M. 2000a. The geology of the Fareham and Portsmouth district — a brief explanation of the geological map Sheet 316 Fareham and part of Sheet 331 Portsmouth. *Sheet Explanation of the British Geological Survey*.

HOPSON, P M. 2000. Geology of the area around North Tidworth, Ludgershall, Netheravon, Tidcombe and Porton Down, Wiltshire and west Hampshire. *British Geological Survey Technical Report*, WA/00/23.

HOPSON, P M. 2001. Geology of the area around South Winchester, Hursley and Braishfield, Hampshire. *British Geological Survey Internal Report*, IR/01/126.

HOPSON, P M. 2001. Geology of the area around Kimbridge, Mottisfont, Houghton and Broughton, Hampshire. *British Geological Survey Internal Report*, IR/01/127.

INSTITUTE OF GEOLOGICAL SCIENCES, and THAMES WATER AUTHORITY. 1978. Hydrogeological map of the south-west Chilterns and the Berkshire and Marlborough Downs including parts of hydrometric areas 39, 42, 43, and 53 (1:100 000). (Dunstable: Waterlow Ltd for IGS.)

JUKES-BROWNE, A J, and HILL, W. 1903. The Cretaceous rocks of Britain. Vol. 2. The Lower and Middle Chalk of England. *Memoir of the Geological Survey of the United Kingdom*.

JUKES-BROWNE, A J, and HILL, W. 1903. The Cretaceous rocks of Britain. Vol. 3. The Upper Chalk of England. *Memoir of the Geological Survey of the United Kingdom*.

KING, C. 1981. The stratigraphy of the London Clay and associated deposits. *Tertiary Research Special Paper*, No. 6, 1–158.

KING, C, and KEMP, D J. 1982. Stratigraphy of the Bracklesham Group in recent exposures near Gosport, Hants. *Tertiary Research*, Vol. 3, 171–187.

MELVILLE, R V, and FRESHNEY, E C. 1982. The Hampshire Basin and adjoining areas, fourth edition, (London: HMSO for Institute of Geological Sciences).

MEYER, C J A. 1871. On the Lower Tertiary deposits recently exposed at Portsmouth. *Quarterly Journal of the Geological Society of London*, Vol. 27, 74–89.

MORTIMORE, R N. 1986a. Stratigraphy of the Upper Cretaceous White Chalk of Sussex. *Proceedings of the Geologists' Association*, Vol. 97, 97–139.

MORTIMORE, R N. 1986b. Controls on Upper Cretaceous sedimentation in the South Downs with particular reference to flint distribution. *In* The Scientific Study of Flint and Chert. SIEVEKING, G de G, and HART, M B (editors). *Proceedings of the Fourth International Flint Symposium* held at Brighton Polytechnic, 10–15 April 1983.

MORTIMORE, R N. 1987. Upper Cretaceous Chalk in the North and South Downs, England: a correlation. *Proceedings of the Geologists' Association*, Vol. 98, 77–86.

MORTIMORE, R N, and POMEROL, B. 1987. Correlation of the Upper Cretaceous Whiite Chalk (Turonian to Campanian) in the Anglo-Paris Basin. *Proceedings of the Geologists' Association*, Vol. 98, 97–143.

MORTIMORE, R N, WOOD, C J, POMEROL, B, and ERNST, G. 1998. Dating the phases of the sub-Hercynian tectonic epoch: Late Cretaceous tectonics and eustatics in the Cretaceous basins of northern Germany compared with the Anglo-Paris Basin. *Zentralblatt für Geologie Palaeontologie*, und 1349–1401; Stuttgart.

OSBORNE WHITE, H J. 1912. The geology of the country around Winchester and Stockbridge. *Memoir of the Geological Survey of Great Britain*, Sheet 299 (England and Wales).

PENN, I E, CHADWICK, R A, HOLLOWAY, S, ROBERTS, G, PHARAOH, T C, ALLSOP, J M, HULBERT, A G, and BURNS, I M. 1987. Principal features of the hydrocarbon prospectivity of the Wessex-Channel Basin, UK. 109–118 in *Petroleum geology of north west Europe*. BROOKS, J, and GLENNIE, K, (editors). (London:Graham Trotman.)

RAWSON, P F, ALLEN, PM, and GALE, A. 2001. The Chalk Group — a revised lithostratigraphy. *Geoscientist*, Vol. 11, No. 1, 21.

REID, C. 1902. The Geology of the country around Ringwood. *Memoir of the Geological Survey of Great Britain*, Sheet 314 (England and Wales).

# 34 SELECTED BIBLIOGRAPHY

REID, C. 1903. Geology of the country around Salisbury. *Memoir of the Geological Survey of Great Britain*, Sheet 298 (England and Wales).

ROBINSON, N D. 1986. Lithostratigraphy of the Chalk Group of the North Downs, southeast England. *Proceedings of the Geologists' Association*, Vol. 97, 141–170.

SIMPSON, I R, GRAVESTOCK, M, HAM, D, LEACH, H, and THOMPSON, H D. 1989. Notes and cross-sections illustrating inversion tectonics in the Wessex Basin. 123–129 in Inversion tectonics. COOPER, M A, and WILLIAMS, G D (editors). *Geological Society of London Special Publication*, No. 44.

SMITH, N. 1985. Structure contours and subcrops of the pre-Permian surface of the United Kingdom (South). *British Geological Survey 150th Anniversary Publication*.

SUMBLER, M G. 1996. British Regional Geology: London and the Thames Valley (fourth edition). (London: HMSO for the British Geological Survey.)

TATTON-BROWN, T in CROOK, J (editor). 1993. *Winchester Cathedral: nine hundred years, 1093–1993*. (Chichester, West Sussex: Phillimore, 1993.)

WHITE, W C F. 1971. A gazetteer of brick and tile works in Hampshire. *Papers and proceedings of the Hampshire Field Club*, Vol. 28, 81–97.

WHITTAKER, A (editor). 1985. *Atlas of onshore sedimentary basins in England and Wales. Post Carboniferous tectonics and stratigraphy*. (Glasgow: Blackie).

WHITTAKER, A. 1910. The water supply of Hampshire. *Memoir of Geological Survey of England and Wales*.

WHITTAKER, A, HOLLIDAY, D W H, and PENN, I E P. 1985. Geophysical logs in British Stratigraphy. *Special Report of the Geological Society, London*, No. 18.

WILKINSON, I P. 2000. Chalk foraminifera from the Broughton–Houghton area of the Winchester Sheet (299). *British Geological Survey Internal Report*, IR/00/31R.

WOODS, M A. 1998a. Review of Upper Cretaceous Chalk (Chalk Group) macrofaunas from the Winchester Sheet (299), Hampshire. *British Geological Survey Technical Report*, WH/98/64R.

WOODS, M A. 1999. Preliminary report on Chalk macrofossils from the Salisbury (298) and Winchester (299) districts. *British Geological Survey Technical Report*, WH/99/88R.

ZIEGLER, P A. 1982. *Geological atlas of western and central Europe*. (Amsterdam: Shell Internationale Petroleum Maatschappij BV.)